Marcelo Monteiro Bezerra
Elisabeth R .A. Cavalcanti Silva
José Gustavo da Silva Melo

Environmental vulnerability on the BR-101 North, Brazil

AF671181

Marcelo Monteiro Bezerra
Elisabeth R .A. Cavalcanti Silva
José Gustavo da Silva Melo

Environmental vulnerability on the BR-101 North, Brazil

Accidents during the transport of dangerous goods

ScienciaScripts

Imprint
Any brand names and product names mentioned in this book are subject to trademark, brand or patent protection and are trademarks or registered trademarks of their respective holders. The use of brand names, product names, common names, trade names, product descriptions etc. even without a particular marking in this work is in no way to be construed to mean that such names may be regarded as unrestricted in respect of trademark and brand protection legislation and could thus be used by anyone.

Cover image: www.ingimage.com

This book is a translation from the original published under ISBN 978-613-9-66949-3.

Publisher:
Sciencia Scripts
is a trademark of
Dodo Books Indian Ocean Ltd. and OmniScriptum S.R.L publishing group

120 High Road, East Finchley, London, N2 9ED, United Kingdom
Str. Armeneasca 28/1, office 1, Chisinau MD-2012, Republic of Moldova, Europe
Printed at: see last page
ISBN: 978-620-8-11552-4

Copyright © Marcelo Monteiro Bezerra, Elisabeth R .A. Cavalcanti Silva, José Gustavo da Silva Melo
Copyright © 2024 Dodo Books Indian Ocean Ltd. and OmniScriptum S.R.L publishing group

DEDICATORY

I dedicate this work to my mother, my brother and my nephew. Also to my father, who I lived with for only 17 years, but who has taught me so that I can live forever.

ACKNOWLEDGEMENTS

I would firstly like to thank God and my family, who have always been there for me on my professional journey. To Professor M.a. Elisabeth Regina Alves Cavalcanti Silva, my supervisor, for giving me the opportunity to present this work. To all the teachers on Fafire's Postgraduate Course in Environmental Expertise and Auditing. To Professor Mércio Buanafina for always replying to my emails and clearing up my doubts, including those relating to land transport.

I cannot fail to express my sincere thanks to everyone at the Pernambuco State Military Fire Brigade, who contributed to the completion of this degree, and also to my mates at IFPE, who serve as a reference in all my work.

Last but not least, I would like to thank my classmates and colleagues in life, especially: Carolina Brito, Liliane Galiza, Emmelyne Luz, Giselli Ervelym, Juvenal Santos, Paulo Roberto, Edjair Xavier and Patrícia Xavier, because without them I wouldn't have succeeded at various stages of this incredible journey. Thank you all very much.

SUMMARY

The general aim of this work is to analyse environmental vulnerability in the face of possible environmental accidents arising from the transport of dangerous products by road on the BR 101 north, specifically in the region around the Capibaribe river. Based on the assessment of the environmental situation of this stretch, together with the specific objectives of identifying the main products transported on the stretch under study, carrying out a prognosis of the possible environmental damage resulting from possible accidents and observing the vulnerable points of the infrastructure of the stretch under study in the event of a possible accident on the south/north and north/south lanes on the stretch of the BR 101 between KM 64 and 65.3 of the BR 101, delimited by the neighbourhood of Dois Irmãos and the DETRAN community. As for the methodology used, the area where the highway crosses the Capibaribe River was characterised by means of a bibliographical analysis. After identifying the structural situation of the road over the River Capibaribe through observations in the field, the problem of the risk of contamination of this river by dangerous products was emphasised, as well as identifying what these products are and what risks to the environment they can cause. The result is that the stretch under study serves as a route for a large quantity of dangerous products coming from various locations, mainly in the south of the country, and this is linked to a prognosis of the possible damage resulting from any environmental accidents based on the concept of risk proposed by Sanchéz (2006). From now on, the environmental vulnerability is notorious due to the lack of supervision of this type of activity and, above all, the absence of preventive measures aimed at reducing the likelihood of accidents involving the transport of dangerous goods by road.

Keywords: Capibaribe River; Hazardous Products; Environmental Risk; Environmental Accidents.

SUMMARY

CHAPTER 1

INTRODUCTION

In recent years, the state of Pernambuco has been undergoing a process of growth in its industrial sector, and with this progress there has been a significant increase in the use of hazardous products, which are necessary for various types of industrial production activities. The main characteristics of these products are the extrapolation of spatial limits, since their action is not restricted to the place where the accident occurred, since these products can, due to their physical state, spread in the form of dust, mist or clouds of contaminants, reaching larger regions than those originally affected, and the extrapolation of temporal limits because their effects can appear several hours, days or years after exposure to the product.

At each stage of the hazardous products production chain, transport requires more attention because it is at this stage that accidents are most likely to occur (ARAÚJO, 2007), in addition to the fact that the transport of hazardous products is a source of diffuse and unpredictable pollution, since it is not possible to know exactly where the polluting substances will be released in the event of a possible accident.

In Brazil, the most widely used means of transporting cargo is land transport, with road transport being the main mode among all the other modes. It is characterised by its simplicity of operation, and stands out for transport over short and medium distances, and for higher value-added cargo (VIEIRA 2001 apud FERREIRA, 2011), but of all the modes it has the lowest load capacity. This mode has a good cost-benefit ratio and also less cargo handling, as it is possible to offer a door-to-door service, i.e. the cargo is taken from one point to another directly (starting point to final destination). However, much of Brazil's road network is in a poor state of repair.

Road transport of dangerous products in the state of Pernambuco is still not given due importance by road and environmental agencies, given the lack of data on accidents, which makes it difficult to know the real risk of this activity in the defined area, since detailed records of these accidents are the basis for knowing the critical roads and stretches in terms of environmental risk.

These facts show us how vulnerable the population and the environment are to the risks of accidents involving the transport of dangerous products on the highways in the state of Pernambuco, especially on the stretches crossed by springs, rivers and lakes, given that these watercourses have the power to dissipate these products even more, and simple mitigating measures such as the installation of signalling, to more complex ones such as modifications to road infrastructure projects are necessary to avoid or reduce negative impacts resulting from possible accidents.

In this sense, the definition of environmental impact adopted in this work, given that it is the most appropriate for the intended objectives, will be that proposed by NBR ISO 14.001:2004, which defines it as any modification of the environment, whether adverse or beneficial, that results in whole or in part from the activities, products or services of an organisation. Therefore, we can see from this concept that the stages of production of dangerous products up to their transport to the end consumer can be causes of environmental changes.

The environmental impacts of accidents involving dangerous chemical products are difficult to measure and can have totally different implications, given that each chemical product released into the environment causes different impacts, and these can vary depending on the type of soil, vegetation and climate in the region where the accident occurred. This is why knowledge of the natural dynamics of the landscape is so important, as it provides parameters that can be used to measure the impacts of anthropogenic activities on the environment (SILVA, 2014).

And when it comes to handling dangerous products, their effects can often go beyond spatial and temporal limits, causing harm, some of it irreversible, to the functioning of a region's natural systems. During the transport of cargo containing dangerous products, various factors can contribute to an abnormal event, such as: human error, mechanical defects in vehicles, poor road maintenance, adverse weather conditions, among others. The impacts related to these types of failures, according to Sánchez (2006), can be more significant than those resulting from the normal operation of a certain activity.

In recent decades, the urbanisation process in Brazilian cities has taken place in a disorderly fashion, causing immense problems in terms of the quality of life of the population and also in terms of structural issues in the country's urban areas. The city of Recife, like other Brazilian metropolises, has not been left out of this chaotic scenario, among other reasons due to the inadequate urbanisation process and the lack of control over the loss of its environmental quality. This scenario can be seen in the example given by the evaluation of the BR-101 northbound motorway in terms of its infrastructure.

This work therefore seeks to diagnose the structural situation of the stretch of the BR-101 motorway around the Capibaribe river in relation to possible accidents in the transport of dangerous goods by road and, in particular, the level of safety where this stretch is located.

CHAPTER 2

OBJECTIVES

2.1 General

To analyse the environmental vulnerability of the stretch of the BR-101 highway in the north of the state of Pernambuco around the Capibaribe river in the face of possible accidents involving the transport of dangerous goods by road.

2.2 Specific:

- Identify the main dangerous products that are transported on the stretch under study;
- Carry out a prognosis of the possible environmental damage resulting from any accidents caused by the transport of dangerous products that may occur in the region;
- Verify critical points to be observed for the possible preparation of a Hazardous Products Emergency Response Plan, specific to the area under study, based on the analysis of accident risks.

CHAPTER 3

LITERATURE REVIEW

3.1 Environmental Impacts

There are many definitions of environmental impact. According to CONAMA Resolution 001 of 23/01/1986, environmental impact is defined as any alteration to the physical, chemical and biological properties of the environment caused by any form of matter or energy resulting from human activities that directly or indirectly affects: I - the health, safety and well-being of the population; II - social and economic activities; III - biota; IV - the aesthetic and sanitary conditions of the environment; and V - the quality of environmental resources.

In order to quantify the environmental impacts resulting from anthropogenic action on the environment, it is necessary to diagnose these impacts. In this way, an environmental assessment presupposes the evaluation of a given site using different methodologies, with the aim of generating an environmental diagnosis, generally describing the levels of contamination, disturbance or preservation of a given area (EMBRAPA, 2010). According to EMBRAPA (2010), making an environmental diagnosis means interpreting the problematic environmental situation of the area, based on the interaction and dynamics of its components, whether related to physical and biological elements or socio-cultural factors.

According to Bastos and Almeida (2006), carrying out an environmental impact assessment generally follows these steps: developing a complete understanding of the proposed action; acquiring technical knowledge of the environment to be affected or studied; determining the possible impacts on environmental characteristics, quantifying the changes where possible; presenting the results of the analysis in such a way that the proposed action can be used in a decision-making process.

3.1.1 Concept of Danger

The concepts of hazards and risk are often seen as synonymous, but there are some subtle differences that are necessary for a correct understanding of these concepts in order to control the logistical management of various activities involving these substances.

The definition used by the Environmental Sanitation Technology Company, CETESB (2003), classifies danger as "one or more conditions, physical or chemical, with the potential to cause harm to people, property, the environment or a combination of these". Another more complete definition, and the one adopted for this work, is given by Morgado (2002), as being: "a capacity or condition inherent in a substance or activity, capable of causing harm to people, property or the environment".

With regard to danger, it is important to note that it is an intrinsic property of a situation, being, material or activity, and because of this characteristic it cannot be controlled or reduced. It is the threat itself attributed to the hazard linked to what the Occupational Health and Safety *Advisory Services,* OHSAS 18.001:2007, defincs as "a source or situation with the potential to cause harm in terms of injury, illness, damage to property, damage to the workplace environment, or a combination of these".

3.1.2 Concept of Risk

The abnormal operation of certain activities can have a negative impact on a given location, and it is from this abnormality that we get the notion of risk. For example, the transport of dangerous goods by road is expected to go according to plan, but if the transport vehicle overturns as a result of a possible accident, we will be faced with what we can call an environmental risk characterised by the possibility of the product being spilled.

Given this, situations associated with a potential for adverse consequences, the results of which entail loss of material goods and damage to human health and the environment, are characterised by what we know as risk (SOCIETY FOR RISK ANALYSIS apud SÁNCHEZ, 2006).

Risk is conceptualised by Sánchez (2006) as the contextualisation of a dangerous situation, i.e. the possibility of the danger materialising or of an undesirable event occurring. And according to Sánchez (2006), this materialisation occurs through the product of the probability of a certain event occurring and the magnitude of the consequences, or

$$R = P \times C$$, where:

R= risk

P= probability of occurrence

C= magnitude of consequences

The definition given by OHSAS 18.001:2007 further emphasises Sánchez's concept, stating that "risk is the combination of the probability of occurrence and the consequence(s) of a given hazardous event". Let's note the mathematical nature of these concepts, making it possible to calculate different risk situations.

The definition of risk given by Fricke (1992) is "the probability of a danger materialising and causing a certain pre-established damage". From this more formal definition we can establish a correspondence between the terms, and at the same time differentiate them so that there is no doubt about their concepts.

3.2 Hazardous Products

There are various definitions of dangerous goods, most of which have common characteristics and specific features; they are the products classified by the United Nations (UN) and published in the Model Regulations - Recommendations for the Transport of Dangerous Goods, known as the Orange Book.

One of the definitions was given in International Labour Convention 174/ILO:1993, as being any substance or mixture of substances which, because of their chemical, physical or toxicological properties, alone or in combination, constitutes a hazard. Compared to the definitions set out below, it can be said that this one addresses the issue in a more generalised way.

The Road Research Institute (IPR), 2011, defines a dangerous product as "any chemical product that poses a risk to people's health, public safety or the environment". In this definition, the public safety factor is one of the key elements, since activities involving dangerous products, if carried out without observing the legal aspects and regulations, can, in the event of accidents, constitute a criminal offence harmful to the environment, which the public safety agencies will be responsible for investigating and reprimanding.

Due to the fact that there are a large number of substances labelled hazardous, many definitions fail to mention one or other characteristic, which can lead to doubts about their classification. Therefore, the definition of dangerous goods as any solid, liquid or gaseous material that is toxic, radioactive, corrosive, chemically reactive or unstable during prolonged storage in a quantity that poses a threat to life, property or the environment (US.DOT, 1998 apud Curso Intervenção em Emergência com Produtos Perigosos. SENASP/MJ.2008) provides a more comprehensive approach.

The properties of these substances are worrying because of their power to cause a series of injuries to living beings and significant impacts on the environment. Hazardous products must therefore be defined on the basis of their common characteristics and specificities, such as: toxic hazards, corrosiveness, chemically reactive or unstable, mechanical hazards caused by explosions, thermal hazards from combustibility, living beings or toxins produced by them, bodies that emit ionising radiation, and which also cause injury, illness or death to those exposed to them, damage to property and the environment.

3.3 Dangerous Goods Transport Regulations

The transport of dangerous goods by road is regulated by Decree 96.044 of May 1988, in which the legislator established the various aspects of the transport activity, such as: cargo packaging; itinerary; documentation; and procedures to be adopted in the event of an emergency. It also determines compliance with the rules of the Ministry of Defence and the National Nuclear Energy Commission (CNEN) when the product to be transported is explosive or radioactive, respectively (BRASIL, 1988).

With regard to the transport of dangerous goods by road in Brazil, there are several government bodies that work together to create the necessary mechanisms for the feasibility of transporting such products. Law 10.233 of 5 June 2001, which restructured the federal transport sector, established in article 22, item VII, that ANTT is responsible for regulating the transport of dangerous cargo and products on roads and railways. Within the scope of ANTT, this activity falls to the Road Freight Transport Regulation Department - GEROC.

The National Institute of Metrology, Standardisation and Industrial Quality (INMETRO) is involved in drawing up technical regulations for the qualification of vehicles and equipment, as well as certifying packaging.

The Ministry of Justice, through the National Transit Council (CONTRAN), regulates issues related to compulsory driver training, and the National Transit Department (DENATRAN) codifies fines related to offences under the National Dangerous Goods Transport Regulations (RTPP) (ARAÚJO, 2007).

The Brazilian Association of Technical Standards (ABNT), on the other hand, works on the development of normative bases for the transport of dangerous products, such as: standards for the identification, handling, movement and storage of products; terminology; standardisation of emergency forms and envelopes for the land transport of dangerous products, their characteristics, dimensions and completion; sets of emergency equipment; among other standards. Some of these standards are voluntary and others are mandatory.

As for the Brazilian Association of Chemical Industries (ABIQUIM), it carries out emergency information and communication work through a service called Pró-Química, serving as a reference for authorities working in emergency response. Its guidance manuals are based on ANTT Resolution 420 of 2004, the UN Guidelines for the Land Transport of Dangerous Goods, 15th edition and the Emergency Response Guidebook from the U.S. Department of Transportation (DOT), Transport Canada (TC), Mexico's Secretariat of Communication and Transport (SCT) and Argentina's Chemical Information Centre for Emergencies (CIQUINE) (ABIQUIM, 2011).

However, for consultation, the main rules and laws relating to the transport of dangerous goods on Brazilian roads are available on the website of the Road Research Institute.

3.4 Basic Emergency and Contingency Concepts

It is during transport, handling and storage that the likelihood of a chemical emergency occurs, which can result in fire or explosion, leakage of toxic gases or vapours, causing serious injuries to people and major impacts on the environment.

The level of difficulty of operations and the various types of accident scenarios will determine the most effective procedures for dealing with an emergency. In order for these procedures to become efficient, even knowing that in an accident one scenario is never the same as another, they must be based on a document called an Emergency Plan or Contingency Plan. According to Araújo (2007), an Emergency Response Plan (ERP) normally involves responding to emergencies whose effects are limited to a company's internal area of operation, whereas when the impacts of an emergency go beyond a company's or agency's area of operation, requiring the involvement of environmental control bodies, federal or state road police, firefighters and civil defence, this plan is called a Contingency Plan.

These plans are essential elements in a Quality, Safety, Environment and Health management system with the aim of triggering quick and effective action. Both the EAP and the Contingency Plan, according to Araújo (2007), should present, in a simple and objective manner, the details involving actions, attributions, responsibilities and communication mechanisms, and should also include some basic principles in their guidelines, such as:

a) Minimise the number of people involved in the contaminated area;
b) Avoid contact with contaminated products and/or surfaces;
c) Keep people informed about control of the emergency site;
d) Provide an escape route and ensure that everyone knows the communication system;
e) Ensure that all aspects related to the level of

responsibility and authority are well defined; among others.

CHAPTER 4

METHODOLOGY

4.1 Characterisation of the Study Area

The area chosen for the study was around the bridge over the River Capibaribe, on the stretch between Km 64 and 65.3 of the BR 101, between the neighbourhood of Dois Irmãos and the DETRAN community, with coordinates 8° 01'38.76" S and 34° 56'34.75" O (figure 1). This stretch is located in the lower reaches of the Capbarbe River, in the northern region of the city of Recife, where there are environmental and social factors vulnerable to the risks characterised by the transport of dangerous goods by road.

Figure 1 - Stretch of the Capibaribe River delimited by the DETRAN and Dois Irmãos communities.
Source: Google Earth (2018).

The study area was then delimited as part of the area of influence of the Capibaribe River, since if it is accidentally affected by a mobile source of pollution, it can quickly reach other localities such as the neighbourhoods of Apipucos, Monteiro, Iputinga and Poço da Panela, which are also affected.

depending on the products and/or their quantity, they can generate acute or chronic risk factors.

The choice was also based on the results of a survey on the most dangerous stretches of the BR 101 carried out by the Federal Highway Police in 2011, published on the Ministry of Justice's website, which analysed the 60 ten-kilometre stretches considered most dangerous by the corporation. According to 2010 data collected by the PRF Department, the stretch between kilometre 60 and 70 of the BR 101 in Pernambuco, which includes the area under study in this paper, ranks twenty-second in terms of accident severity, with 9 (nine) fatalities, 200 (two hundred) injured victims and 518 (five hundred and eighteen) accidents without victims. And with the future construction of a river transport terminal under the Pernambuco state government's urban mobility plan, the tendency is for there to be a greater flow of vehicles and people in this area, although the work has been at a standstill since last year.

Another important factor that has been taken into consideration is the mangrove areas located on the banks of the Capibaribe River in this stretch, which are highly productive and complex, because according to Schaeffer-Novelli (2001), mangroves have a very strong functional variety and, consequently, tend to be more resistant to both natural and human-induced disturbances, but each subsequent disturbance causes damage to the system, making them more vulnerable to these exposures. And depending on the magnitude of the events involving hazardous products, the environmental consequences for this system can be severe.

4.2 Characterisation of the object of study

The BR 101 is a longitudinal motorway that runs from north to south along the entire eastern coast of Brazil (DNIT, 2008). Among the 70 kilometres that make up the Recife Metropolitan Region, there is a stretch that crosses the Capibaribe basin, more specifically the lower course of the river in the northern region of Recife. This is a worrying factor, given the possible impacts caused to the environment by the excessive flow of vehicles and, therefore, in accordance with the legal provisions inherent to the environment, any environmental diagnosis or analysis of environmental impacts on water resources must refer to the river basin in which it is located (SILVA, 2014).

The area of influence of the watercourse at the point where the BR 101 over the River Capibaribe is located covers the neighbourhoods of Apipucos, Monteiro, Iputinga, Poço da Panela, Santana, Torre and the Detran community, totalling 5 km in length. This will take into account the agents acting or omitted that could directly or indirectly interfere in the dynamics of possible contamination of the Capibaribe River by contact with chemicals considered dangerous.

4.3 Classification of Dangerous Goods

Brazil adopts the classification of dangerous products accepted internationally by the member countries of UNEP (United Nations Environment Programme), regulated by Decree 96.044/1988 (Regulation on the Transport of Dangerous Products - RTPP), whose complementary instructions were approved by Resolution 420/2004 of the National Land Transport Agency (ANTT) and amended by Resolution 701/2004.

The aforementioned Resolution divides dangerous products into nine classes, which in turn are subdivided into subclasses, according to the risk or most severe risk they present, as shown in Table 1.

Table 1. Classes and Subclasses of dangerous products

CLASS AND SUBCLASSES	NAME
Class 1	**Explosive**
Subclass 1.1	Substances and artefacts with a risk of mass explosion.
Subclass 1.2	Substances and articles with a projection risk but no mass explosion risk.
Subclass 1.3	Substances and articles with a risk of fire and a small risk of explosion or projection, or both, but no risk of mass explosion.
Subclass 1.4	Substances and articles that do not present a significant risk.
Subclass 1.5	Very insensitive substances, with a risk of mass explosion.
Subclass 1.6	Extremely insensitive articles, with no risk of mass explosion.
Class 2	**Gases**
Subclass 2.1	Flammable gases.
Subclass 2.2	Non-flammable, non-toxic gases.
Subclass 2.3	Toxic gases.
Class 3	**Flammable liquids**
Class 4	**Flammable solids; substances subject to spontaneous combustion; substances that emit flammable gases in contact with water.**
Subclass 4.1	Flammable solids, self-reacting substances and insensitised solid explosives.
Subclass 4.2	Substances subject to spontaneous combustion.
Subclass 4.3	Substances that emit flammable gases when in contact with water.
Class 5	**Oxidising Substances and Organic Peroxides**
Subclass 5.1	Oxidising substances
Subclass 5.2	Organic Peroxides
Class 6	**Toxic Substances and Infectious Substances**
Subclass 6.1	Toxic (poisonous) substances

Subclass 6.2	Infectious substances
Class 7	**Radioactive Material**
Class 8	**Corrosive substances**
Class 9	**Miscellaneous Hazardous Substances and Articles**

Source: ABIQUIM (2011).

4.4 Identification of Hazardous Products

According to the Hazardous Products Emergency Response Manual of the Brazilian Chemical Industries Association (ABIQUIM, 2011), the product hazard identification codes are the numbers seen at the top of the safety panel (orange plate) of vehicles transporting hazardous products. Each code is made up of two or three digits and, in certain cases, preceded by the letter "X" when they react violently with water.

In the 2004 Resolution of the National Land Transport Agency, ANTT 420, these codes are called Risk Number. The digits that make up the code are associated with the hazardous properties of the products, articles and waste being transported, as shown below:

2 - Gas release due to pressure or chemical reaction;

3 - Flammability of liquids (vapours) and gases or liquid subject to self-heating;

4 -Inflammability of solids or solids subject to self-heating;

5 - Oxidising effect (intensifies the fire);

6 - Toxicity or risk of infection;

7 - Radioactivity;

8 - Corrosivity;

9 - Spontaneous violent reaction and other hazards not covered by the above figures; and

X - Dangerous reaction with water.

Complementary notes in the ABIQUIM Manual for Responding to Emergencies with Dangerous Products, 2011:

a) When the hazard associated with a substance can be adequately indicated by a single digit, this will be followed by zero; for example: 30- Flammable Liquid.

b) The repetition of a digit indicates greater intensity of a specific risk; for example: 33-Highly flammable liquid.

c) If the risk number is preceded by the letter "X" it means that the product reacts dangerously with water.

d) Codes that include "9" as a 2° or 3° digit (e.g. 59, 339): indicate danger of violent spontaneous reaction, whether explosion, violent disintegration or (explosive) polymerisation reaction. Does not apply to "99".

e) Codes beginning with "9" (i.e. 90 or 99): used only for products, articles and hazardous waste that do not have the dangerous properties identified by the numbers 2 to 8. This includes, for example, products that are dangerous for the environment, genetically modified micro-organisms and products transported at high temperatures.

4.5 Structure of a Service Plan

According to the roadmap for drawing up an Emergency Action Plan made available by the Ministry of the Environment on its website, the plans must have an organised structure and be a practical tool for rapid and effective responses in emergency situations:

1- Introduction

- Provide a brief history of accidents involving chemical products, emergency response activities and the availability of infrastructure.

2- Objectives

- Establish technical and administrative procedures to be adopted in emergency situations in the region;
- Promote basic measures to restrict damage to a previously dimensioned area, in order to prevent impacts from exceeding pre-established safety limits;

- Indicate the actions aimed at avoiding impacts and those that may contribute to aggravating them;
- Clearly and objectively define the roles and responsibilities of those involved.

3- Definitions

- Explanation of the main technical terms used.

4- Characterisation of the area

- Description of existing segments and facilities and surrounding population densities, aspects of use and occupation and proximity to vulnerable environmental areas.

5- Basic assumptions

- Considerations, justifications and reasons for need.

6- Area covered by the plan

- Location and area - regional, municipal, state or federal.

7- Accidental hypotheses

- Description of the areas where accidents may occur or emergency activities may take place.

8- Examples of accidents

- Types of accidents and expected consequences in each accidental hypothesis considered, with the impacts on vulnerable areas in the region.

9- Organisational structure

- Organisational chart with a schematic presentation of the plan's organisational structure, coordination, working groups and teams;
- Duties and responsibilities of coordination, working groups and teams, with a description of the activities and obligations of those involved.

10-Drive

- EAP triggering flowchart showing the sequence of triggering steps and the hierarchical decision-making level of those involved.

11-Emergency procedures

-Assessment and identification of the problem, size of the event and initial procedures to control the situation;

-Control procedures : a) actions to combat emergencies and measures to minimise their consequences and impacts - size, type of occurrence, jurisdiction and attributions of participants; b) isolation; c) stoppage of activities; d) evacuation of people; e) firefighting; f) control of leaks; g) emergency repairs; h) rescue; i) treatment of poisoned people;

Post-emergency (aftermath) actions to re-establish normal conditions in the areas affected by the consequences of the accident.

12-Human and material resources

- Planning and compatibility with the size of the expected occurrences and sizing to subsidise the technical and operational needs established in the control procedures.

13-Training

- Training of plan participants, through individual or collective training to maintain and operationalise work routines;

- Simulation in the field, to train the teams in the procedures and actions to combat accidental episodes.

14- Updating, evaluation and maintenance

- The plan must have: a) a system for reviewing, maintaining and updating it on an ongoing basis, in line with the experience gained from both the services provided and specific training, and performance gauges to assess the efficiency and effectiveness of the targets and objectives set; b) a system for updating information; c) a record of services provided; d) periodic reassessment of procedures; e) replacement and renewal of human and material resources.

15- Disclosure

- Distribution of information about the Plan to participants, public and private sectors.
- rivates, with an interest or link in the development of the activities;
- Integration with other plans;
- The plan should provide for integrated work with other plans, involving public and private institutions and/or mutual aid organisations in a given locality, as well as the involvement and participation of the National P2R2 Commission.

16- Annexes

- Occurrence recording forms, reports and telephone answering forms;
- List of organisations to contact and list of emergency telephone numbers;
- Protocol and work instructions, procedures, competency requirements,

Material Safety Data Sheet (MSDS);

- List of human and material resources;
- List of technical teams, companies, public bodies, available material resources (machinery, personal protective equipment, environmental monitoring, leak prevention and containment, etc.), among others;
- Bibliographical references: municipal, state and federal legislation, tables, specific laws, regional prohibitions, compulsory licences, technical standards, etc.

CHAPTER 5

RESULTS AND DISCUSSION

5.1 Environmental vulnerability around the Capibaribe River

This work is based on the Risk Analysis Study, which according to Sánchez (2006) is required for the licensing of certain industries or other potentially dangerous activities, such as the transport of dangerous goods by road. According to Sánchez (2006), in addition to identifying the dangers and estimating the risks, such studies must also propose management measures.

The results of this research are based on comparisons made between what was found in field research and certain items in the Manual for the Implementation of Emergency Action Plans for Responding to Accidents Involving the Road Transport of Dangerous Goods drawn up by the DNIT.

As well as being a basic orientation tool, as a guide for implementing emergency action plans for immediate responses to accidental events involving dangerous goods on highways (DNIT, 2005), this manual also offers instruments for adapting the physical infrastructure on highways in order to minimise the risks of accidents involving the transport of dangerous goods.

It is important to emphasise that the observations were made only through photographic records in the field. Neither stakes nor single line drawings were used in positions with local references and GPS annotations of the coordinates of the observed locations, as the Manual directs in the phase of surveying the operational conditions of the track, due to the lack of such instruments.

5.2 Hazardous Products Transported on the Stretch Under Study

As previously mentioned, the identification of hazardous products whose route is along the stretch of the BR 101 under study is done through the website of the Road Research Institute, where, in order to comply with the Regulations on the Transport of Hazardous Products, article 10 of Decree 96.044, all suppliers and distributors must use this means of communication to inform their annual transport. This stretch of road is emphasised in this work because it is the most vulnerable to accidents involving dangerous products.

Registration is done as follows: 1. After entering the IPR website (http://ipr.dnit.gov.br), in the main menu on the left of the screen, click on dangerous goods; 2. Click on route registration; 3. On the next screen, the icon on the left has the option to register the company and users. These must be filled in according to the sequence of data requested for the products the company works with; 4. Once all the fields have been filled in and sent, the data will be validated by the National Department of Transport Infrastructure/Institute of Road Research, DNIT/IPR. This validation normally takes place on weekdays.

Once validated by the DNIT, the day after registration, the user can access "route registration" with their login and password, and a screen will open with the name of the company validated and recognised by the system. From there, the applicant must enter the UN number of the product, the Federation Unit and Municipality of origin, the Federation Unit and Municipality of destination, the tonnage transported, and the roads travelled from Origin to Destination.

If the road is "recognised" by its regional name, the system offers the option of identifying the corresponding acronym; click on the column header and locate the road acronym to replace the regional name. Each screen to be filled in is for a single product, i.e. if the product has several destinations, as many lines per screen as there are destinations must be filled in. In the quantity column, the two units must be filled in: t or m^3 , for example: t = 3; m^3 = 0. If they are not filled in, the system will not accept the product registration. Finally, once the data is consistent and sent, you can request to view the report of the route records entered during the period on the IPR/DNIT web system.

The following table shows the products transported in 2010 and registered in 2011. The search can be carried out by year and risk class or year and product, but due to the large number of products registered on the IPR website, the attributes used for this study were year and risk class.

Some of the products in tables 1 and 2 had their names recorded more than once, but what was emphasised was the route used to transport them. Another aspect that should be noted is that in the "road" attribute, the routes are not in order of the journey made between the origin and destination of the product, always emphasising the BR 101.

Table 1. Register of routes used to transport dangerous products.

Gases - Class 2							
Registration Date	**Refers to the year**	**Product**	UN NO.	**Risk No.**	**Origin**	**Destination**	**Motorway**
9/2/2011	2010	Chlorine (SC-2.3)	1017	268	PE- Itapissuma	PE- Afogados da Ingazeira	BR 101;232;110
9/2/2011	2010	Chlorine (SC-2.3)	1017	268	PE- Itapissuma	PE-Agua Preta	BR 101
9/2/2011	2010	Chlorine (SC-2.3)	1017	268	PE- Itapissuma	PE- Barreiros	BR 101
9/2/2011	2010	Chlorine (SC-2.3)	1017	268	PE- Itapissuma	PE-Caruaru	BR 101; 232
9/2/2011	2010	Chlorine (SC-2.3)	1017	268	PE- Itapissuma	PE-Ladder	BR 101
9/2/2011	2010	Chlorine (SC-2.3)0	1017	268	PE- Itapissuma	PE-Recife	BR 101
11/4/2011	2010	Nitrogen Peroxide (SC- 2.3)	1067		PE-Jaboatão dos Guararapes	PE- Paulista	BR 101 North
11/4/2011	2010	Nitrogen Peroxide (SC-2.3)	1067		PE-Jaboatão dos Guararapes	PE-Olinda	BR 101 North
11/4/2011	2010	Nitrogen Peroxide (SC- 2.3)(*)	1067		PE-Jaboatão dos Guararapes	PE-Recife	BR 101 North

Source: Adapted by the authors (2018).

Table 2. Register of routes used to transport flammable liquids.

Flammable Liquids - Class 3							
Registration Date	**Referring to the Year**	**Product**	**UN NO.**	**Risk No.**	**Origin**	**Destination**	**Motorway**
24/3/2011	2010	Ethyl Acetate(*)	1173		SP- Guarulhos	PE-Recife	BR 101;116;381; 324
29/3/2011	2010	Hydrocarbon(s) Liquid(s); N.E(').	3295	30	SP- Guaratinguetá	PE-Recife	BR 101;040;116; 324;316
30/3/2011	2010	Diesel oil	1202	90	RN-Natal	PE-Ipojuca	BR 101; PE 060
31/3/2011	2010	Ethyl Glycol	1171		SP- Guarulhos	PE-Olinda	BR 101;116;381; 324
1/4/2011	2010	Vinyl Acetate, Stabilised(*)	1301		SP-Santos	PE-Recife	BR 101,116,393, 342; SP 150,116; PE 008
1/4/2011	2010	Butyl Acrylate(s), Stabilised(*)	2348		PE-Ipojuca	PE-Recife	BR 101,232,316, 230,343
1/4/2011	2010	Butyl Acrylate(s), Stabilised	2348		PE-Ipojuca	PI- Teresina	BR 101; PE 008
4/4/2011	2010	Liquid Flammable, N.E.(*)	1993		SP- Guaratinguetá	PE-Recife	BR 101,116,324
4/4/2011	2010	Resin, Solution	1866	30	PE-Jaboatão dos Guararapes	RN-Natal	BR 101,232,230
7/4/2011	2010	Paint and flammable materials (including paint thinners and reducers)	1263	30	PE-Jaboatão dos Guararapes	RN-Natal	BR 101,232,230
8/4/2011	2010	Flammable paints	1263	30	PE-Jaboatão	RN-Natal	BR 101, 232,230

		(including lacquers, enamels, dyes, shellacs, varnishes, polishes, liquid fillers and liquid bases for lacquers.			dos Guararapes		
11/4/2011	2010	Methanol	1230		PE-Jaboatão dos Guararapes	PE-Olinda	BR 101 North
14/4/2011	2010	Xylene(*)	1307		SP- Guarulhos	PE-Recife	BR 101,116,324, 381
18/4/2011	2010	Stickers Containing Flammable Liquids	1133		MG-Lagoa Santa	PE-Abreu e Lima	BR 101,116,259; PE 15;LMG 800

Source: Adapted by the authors (2018).

Table 3. Flammable Solids, Self-Reacting Substances and Insensitised Solid Explosives - Class 4.

Registration Date	Referring to the Year of:	Product	UN NO.	Risk No.	Origin	Destination	Roads
29/3/2011	2010	Self-heating solid, organic, N.E. (SC-4.2)(')	3088	40	SP-Guarulhos	PE- Carpina	BR 101,116,262, 381;VI999
30/3/2011	2010	Sulphur (SC-4.1)(*)	1350		SP-Guarulhos	PE-Recife	BR 101,116,381, 324

Source: Adapted by the authors (2018).

Table 4. Oxidising Substances and Organic Peroxides - Class 5.

Registration Date	Referring to the Year of:	Product	UN NO.	Risk No.	Origin	Destination	Roads
24/1/2011	2010	Sodium Chlorate (SC-5.1)	1495	50	GO- Anápolis	PE-Itapissuma	BR 101,116,06 0, 020,242,34 2, 407
31/1/2011	2010	Calcium hypochlorite, dry or mixture of dry calcium hypochlorite with more than 39% free chlorine (SC-5.1)	1748	50	PE-Itapissuma	PE-Cabo de Santo Agostinho	BR 101
31/1/2011	2010	Calcium hypochlorite, dry or mixture of dry calcium hypochlorite with more than 39% free chlorine (SC-5.1)	1748	50	PE-Itapissuma	PE- Custody	BR 101, 232
15/3/2011	2010	Anhydrous Chromium Trioxide (SC-5.1)(*)	1463	58	SP-São Paulo	PE-Recife	BR 101, 116
22/3/2011	2010	Solid Oxidising, Corrosive N.E. (SC-5.1)(*)	3085	58	SP-São Paulo	PE-Carpina	BR 101, 116
27/3/2011	2010	Hydrogen Peroxide, Aqueous Solution containing not less than 20% but not more than 60% of Oxygen Peroxide	2014	58	ES-Aracruz	PE-Paulista	BR 101, 324; ES 010,445

		(SC- 5.1)					

Source: Adapted by the authors (2018).

Table 5. Oxidising Substances and Organic Peroxides - Class 6.

Registration Date	Referring to the Year of:	Product	UN NO.	Risk No.	Origin	Destination	Roads
22/3/2011	2010	Inorganic Toxic Liquid, N.E. (SC-6.1)(*)	3287	60	SP-São Paulo	PE-Recife	BR 101,116
31/3/2011	2010	Toxic Solid Pesticide, N.E. (SC-6.1)(*)	2588	60	SP-Guaratinguetá	PE-Recife	BR 101,116, 040,267,324, 316;SP 060; RJ159,151, 147; MG 353; AL 110,102
1/4/2011	2010	Toluene diisocyanate (SC-6.1)	2078		BA- Camaçari	PB-João Pessoa	BR 101,110; BA 093
1/4/2011	2010	Toluene diisocyanate (SC-6.1)	2078		BA- Camaçari	PE-Olinda	BR 101
4/4/2011	2010	Trichloroethylene (SC-6.1)	1710		SP-Guarulhos	PB-João Pessoa	BR 101, 381,324,116
4/4/2011	2010	Toxic Liquid Pesticide, N.E. (SC-6.1)(*)	2902	60	SP-Guaratinguetá	PE-Recife	BR 101, 040,267,116, 324,316; SP 060; MG 353; RJ 159,151,147; AL 110,102
5/4/2011	2010	Organochlorine-based Pesticide, Liquid, Toxic, Flammable, with PFG Equal to or Above 23° C (SC-6.1)(*)	3017	63	SP-Guaratinguetá	PE-Recife	BR 101,040,116, 267,324,316; SP 060; RJ159,151, 147; MG 353; AL 110,102
5/4/2011	2010	Pesticide based on Organophosphorus, Solid, Toxic (SC-6.1)(*)	2783	60	SP-Guaratinguetá	PE-Recife	BR 101,040,116, 267,324,316; SP 060; RJ159,151, 147; MG 353; AL 110,103
11/4/2011	2010	Perchloroethylene (SC-6.1)(*)	1897		SP-Guarulhos	PE-Recife	BR 101, 116,381,324

Source: Adapted by the authors (2018).

Table 6. Corrosive substances - Class 8.

Registration Date	Referring to the year	Product	UN NO.	Risk No.	Origin	Destination	Roads
20/10/2010	2010	Phosphoric Acid, Liquid(*)	1805	80	SP-São Paulo	PE-Recife	BR 101,116, 381,040, 135,324, 251; SP 015
24/1/2011	2010	Sulphuric Acid, Solution with Up to 51% Acid	2796	80	GO-Anápolis	PE-Itapissuma	BR 101,116, 342,242, 407,060,020
29/1/2011	2010	Corrosive liquid, N.E.	1760	80	PE- Itapissuma	PE-Cabo de Santo Agostinho	BR 101
31/1/2011	2010	Hypochlorite, Solution	1791	80	PE- Itapissuma	PE-Jaboatão	BR 101

						dos Guararapes	
1/3/2011	2010	Phosphoric Acid	1805	80	SP-São Paulo	PE-Olinda	BR 101,116, 381,040, 135,251,324
4/3/2011	2010	Liquid Alkylsulfonic Acid(s), with up to 5% Free Sulfuric Acid	2586	80	SP-São Paulo	PE-Olinda	BR 101,116, 381,040, 135,251, 325; SP 015
9/3/2011	2010	Thioglycolic acid(*)	1940	80	SP-São Paulo	PE-Recife	BR 101,116, 381,040, 135,251, 325; SP 016
28/3/2011	2010	Acetic Acid, Glacial(')	2789		SP-Guarulhos	PE-Recife	BR 101,116, 324,381
1/4/2011	2010	Sodium Hydroxide, Solution	1824		BA-Candeias	PE-Igarassu	BR 101,324; BA 522
4/4/2011	2010	Mixture of Hydrofluoric Acid and Sulphuric Acid(')	1786		SP-Barueri	PE-Recife	BR 101,116; SP 312,280
4/4/2011	2010	Monoethanolamine	2491		SP-Guarulhos	PE-Olinda	BR 101, 116, 324,381
5/4/2011	2010	Corrosive Liquid, Basic, Organic, N.E.	3267	80	SP-Guaratinguetá	PE-Abreu e Lima	BR 101,116, 393,267, 420; RJ257,145, 151; VI999
26/4/2011	2010	Isoforonadiamine(*)	2289		SP-Taboão da Serra	PE-Recife	BR 101,116, 381; SP021
12/7/2011	2010	Maleic Anhydride(*)	2215	80	SP-Várzea Paulista	PE-Recife	BR 101,116; SP354,065

Source: Adapted by the authors (2018).

Table 7. Miscellaneous Hazardous Substances and Articles - Class 9.

Registration Date	Referring to the Year of:	Product	UN NO.	Risk No.	Origin	Destination	Roads
30/3/2011	2010	Granular, Expandable Polymers that Release Flammable Vapours	2211	90	SP-Guaratinguetá	PE-Olinda	BR 101,116,040, 267,316,324; RJ159,151, 147; MG 353; AL 110,102; PE 015
1/4/2011	2010	Substance Presenting a Risk to the Environment, Liquid, N.E.	3082		SP-Jundiaí	PE-Paulista	BR 101,116,324, 393; SP330,063,065
6/4/2011	2010	Substance Posing a Risk to the Environment, Solid, N.E. (*)	3077	90	SP-Guaratinguetá	PE-Recife	BR 101,116,040, 267,316,324; RJ159,151, 147; MG 353; AL 110,102

(*) Products which, according to the information, there is no certainty about passing through the stretch under study.

Source: Adapted by the authors (2018). http://ipr.dnit.gov.br/, 2011

5.3 Prognosis of Possible Damage from Transport Accidents

The damage caused to the environment by each of these products is listed in their Chemical Safety Data Sheets (MSDS). These documents are drawn up by the manufacturers of hazardous chemical products in order to inform their users of the dangers of a chemical product, enabling them to act in accordance with a risk assessment, taking into account the conditions of use of the product, and to take the necessary measures to develop an active safety, health and environmental programme. This last aspect is a relevant factor, given that the area under study is already suffering from anthropogenic action, since according to Melo et al. (2011), in this area there is a reduction in the tree composition of the mangrove in the lower reaches of the Capibaribe River as a result of the destabilisation of the sediment, caused by the silting up of the river, or by activities to destructure the mangrove, such as cutting down trees, dumping rubbish, landfills for the construction of homes, among other actions.

The MSDS is a document that, according to Decree No. 2,657 of 03/07/1998, must be received by employers who use hazardous chemical products, via suppliers and manufacturers. In Brazil, the MSDS model is currently defined by ABNT's Brazilian Standard (NBR) 14.725, the second edition of which was published in 2005.

This NBR, which was based on ISO 11.014:1994, aims to provide information on how to prepare and fill in an MSDS, and defines that the document must contain 16 mandatory sections, with the following numbering and sequence:

1 - Product and company identification;
2 - Composition and information on ingredients;
3 - Hazard identification;
4 - First aid measures;
5 - Fire-fighting measures;
6 - Control measures for spillages or leaks;
7 - Handling and storage;
8 - Exposure control and personal protection;
9 - Physico-chemical properties;
10 - Stability and reactivity;
11 - Toxicological information;
12 - Ecological information;
13 - Treatment and disposal considerations;
14 - Transport information;

15 - Regulations; and

16 - Other information.

Given this knowledge, the items worth highlighting in this section are 03, 11 and 12 of the MSDSs for each product registered on the route under study, hazard identification, toxicological information and ecological information, respectively. They present the damage caused to humans, animals and the ecosystem in general, especially in the aquatic environment, at different concentrations and in different exposure routes.

The items that will be presented are from the data sheets of different suppliers and manufacturers, due to the large quantity and variety of the products listed. Below are tables with the main toxicological and ecological information common to products of the same class registered with the IPR, separated by risk class.

1- Gases (class 2)

Table 2. Main toxicological and ecological information for class 2 products.

Products	**- Chlorine** **- Nitrogen Dioxide**
Main Toxicological Information	- Overexposure can cause irritation of the mucous membranes, sinuses, pharyngitis and bronchitis with pain, headache, cyanosis, irregular breathing, choking, dizziness and the possibility of pulmonary oedema. - High concentrations of vapour can cause pain, choking, bronchoconstriction, a slow heart reflex and the possibility of asphyxiation. Lack of oxygen can cause death.
Main Ecological Information	- These products are not considered a marine pollutant by the DOT.

Source: Product MSDS from White Martins Gases Industriais S.A. and Braskem S.A. (2018).

2- Flammable liquids (class 3)

Table 3. Main toxicological and ecological information for class 3 products.

Products	**- Flammable liquids N.E.** **- Butyl Acrylate** **- Stickers** **- Paints** **- Hydrated Ethyl Alcohol Fuel** **- Biodiesel** **- Ethylene glycol** **- Vinyl and Ethyl Acetate** **- Resins Flammable Solutions** **- Liquid Hydrocarbon** **- Methanol**

Main Toxicological Information	**- By inhalation:** Acute: The high concentration of aliphatic hydrocarbon solvents contained in the product may cause dizziness, headache, nausea and mental confusion. Chronic: Aromatic hydrocarbons, present in up to 17% of mineral spirits, can cause chronic damage to the kidneys, liver and central nervous system. Chronic inhalation of dust from sanding the product can cause lung damage. **- By ingestion:** Acute: Nausea, vomiting, diarrhoea; aspiration of vomited contents can cause pneumonitis. Chronic: Not expected at occupational level, nor reported for aliphatic hydrocarbons;
	but the presence of aromatic hydrocarbons in turpentine could potentially lead to liver damage. **- Cutaneous (skin):** Acute: Redness, itching. Chronic: Dryness, eczematous dermatitis. **- Ocular (eyes):** Acute: Irritation of mucous membranes; corneal damage may occur. Chronic: No data available.
Main Ecological Information	- It can affect the soil and, through percolation, degrade water quality in the groundwater. - The environmental effects are the contamination of watercourses, making them unfit for use for any purpose and potentially destroying the fauna and flora at the site of the spill. - Run-off into the sewage system can create fire and explosion hazards.

Source: MSDS of products from the companies M-I Swaco do Brasil - Comércio, Serviço e Mineração Ltda; Bandeira Química Ltda; Petrobras Distribuidora S.A.; Renner Sayerlack S/A; Casquímica Produtos Químicos Ltda; Formiline Indústria de Laminados Ltda; Companhia Sulamericana de Tintas e Solventes; Tec-screen Indústria de Produtos Técnicos para Serigrafia Ltda (2018).

3- Flammable solids (class 4)

Table 4. Main toxicological and ecological information for class 4 products.

Product	**- Sulphur**
Main Toxicological Information	**- By inhalation:** irritation of the upper airways. **- On contact with the skin:** local irritation occurs in some individuals which can be aggravated by perspiration or humidity. **- In direct contact with the eyes:** irritation of the conjunctivae. **- By ingestion:** they can be aspirated into the lungs and cause chemical pneumonia.
Main Ecological Information	**- Environmental impact:** The gases resulting from burning are highly polluting. **- Effects on aquatic organisms:** Fire control wastewater and dilution water can cause pollution. In solid form it is insoluble. **- Effects on soil organisms:** can be slowly oxidised to SO2 by microorganisms, and the rate of oxidation is controlled by the temperature, humidity

	and aeration of the soil.

Source: Product MSDS from Bandeirante Brazmo Indústria e Comércio Ltda (2018).

4- Oxidising substances (class 5)

Table 5. Main toxicological and ecological information for class 5 products

Products	**- Sodium Chlorate, Crystal** **- Calcium hypochlorite** **- Hydrogen Peroxide** **- Solid Oxidiser** **- Chromium Trioxide, Anhydrous**
Main Toxicological Information	- **By inhalation:** can cause irritation to the nose, mouth, throat and lungs, if exposure is chronic, impairment of lung function and permanent damage to the lungs can occur; - **By contact:** can cause severe irritation and burns to the skin, repeated exposure can cause tissue destruction due to the corrosive nature of the product; In contact with the eyes can cause damage to the eyesight and cornea; - **By ingestion:** can cause irritation and/or burns in the gastrointestinal tract, cause haemorrhages and ulceration of tissues.
Main Ecological Information	- Toxic to aquatic organisms, they can cause long-term adverse effects in the aquatic environment.

Source: MSDS of the products by the companies Canexus Brasil Ltda; Mazzarollo Indústria Química Ltda; Bandeirante Química Ltda; Indústria Química ZequiniLtda; Domingos Araújo Neto Corantes e Produtos Químicos (2018).

5- Toxic and Infectious Substances (class 6)

Table 6. Main toxicological and ecological information for class 6 products.

Products	**- Toluene diisocyanate** **- Toxic Liquid, N.E.** **- Perchloroethylene** **- Organochlorine-based, Liquid, Toxic, Flammable Pesticides with pfg = or > 23° C** **- Organophosphate-based Pesticides, Solid** **- Liquid Pesticides, Toxic, N.E.** **- Solid, Toxic Pesticides, N.E.** **- Trichloroethylene**
Main Toxicological Information	**- Adverse effects on human health:** the products are irritating to the skin, eyes and mucous membranes and can lead to neurological symptoms.
Main Ecological Information	- Harmful to aquatic organisms and may cause long-term adverse effects in the aquatic environment. - Highly toxic to birds and insects.

Source: MSDS of products from the companies BandeiranteBrazmo Indústria e Comércio Ltda; Electrochemical Produtos e Processos Galvânicos; Quimidrol Comércio Indústria Importação Ltda; Milenia Agrociências S/A; ArystaLifeScience do Brasil Indústria Química e Agropecuária Ltda; Basf S.A (2018).

6- Corrosive substances (class 8)

Table 7. Main toxicological and ecological information for class 8 products.

Products	**- Acetic Acid, Glacial** **- Alkylsulfonic Acid, Liquid with up to 5% Free Sulfuric Acid**

	- Liquid Phosphoric Acid **- Sulphuric acid, with more than 51% acid** **- Thioglycolic acid** **- Oxalic acid** **- Malic Anhydride** **- Anhydrous Hydrofluoric Acid** **- Sodium hydroxide** **- Sodium hypochlorite** **- Isoforonadiamine** **- Corrosive Liquid, Basic, Organic, N.E.** **- Monoethanolamine** **- Aluminium polychloride**
Main Toxicological Information	**- Acute Toxicity** **Inhalation:** Causes irritation of the respiratory tract and causes coughing. In more serious cases can cause chemical burns **Ingestion:** Ingestion can cause burning pain due to the presence of sulphuric acid, causing abdominal pain and vomiting. **Skin:** May cause irritation and reddening (sensitisation) of the skin. In more serious cases, they can cause burns **Eyes:** May cause irritation and burns. **Acute local effects:** The severity of the accident (contact) with the product depends on the sensitivity of the individual, the time of contact and the concentration. **- Chronic toxicity** **Systemic effects:** Repeated and prolonged contact can cause irritation, dermatitis and burns. **Main symptoms:**May cause irritation and reddening (sensitisation) of the skin. In more serious cases, they can cause burns.
Main Ecological Information	- Products classified as dangerous for aquatic organisms. - Products can cause damage to the environment due to their corrosiveness and acidity. - Discharge into the environment should be avoided.

Source: MSDS of the products by the companiesElectrochemical Produtos e Processos Galvanotécnicos; Morais de Castro Comércio Importação de Produtos Químicos Ltda; Alvasi Comercial Ltda; Elekeiroz S.A.; Merck S.A. do Brasil; Petrom - Petroquímica Mogi das Cruzes S.A.; Companhia Nitro Química Brasileira; Bandeirante Química Ltda; Denver Impermeabilizantes Indústria e Comércio Ltda; Quimiclor Comercial Ltda; Brasinter Produtos Químicos Ltda; Superquímica Comércio e Transporte Ltda (2018).

7- Miscellaneous Hazardous Substances and Articles (class 9)

Table 8. Main toxicological and ecological information for class 9 products.

Products	**- Granulated, Expandable Polymers** **- Substances Posing a Risk to the Environment, Liquid, N.E. (pelletised carbon)** **- Substance posing a risk to the environment, solid, N.E. (chlorothalonil)**
Main Toxicological Information	**Local effects:** **Inhalation:** Causes airway irritation superior.

	Skin contact: May cause mild to moderate irritation. Prolonged and repeated skin contact can be dangerous. **Eye contact:** No prolonged or significant irritation is expected.
Main Ecological Information	- They cause visual pollution (pearls). - Leaks and spillages can cause mortality of aquatic organisms and damage to terrestrial flora and fauna. - They can impart undesirable qualities to the water, affecting its use.

Source: MSDS of the products from the companies Denver Impermeabilizantes Indústria e Comércio Ltda; Ultrafértil S.A.; Iharabras S.A. Indústrias Químicas (2018).

5.4 Observation of the Structural Points of the Stretch Under Study

Starting by observing the stretch of the BR 101 motorway around the Capibaribe River, in both directions of the stretch delimited in the research, we can see that in the south/north direction it is of great strategic importance for the transport of dangerous products, as the south-east is the region of origin for 63% of the supplier and distributor companies (figure 2), given that of the 46 products listed, 41 originate in the south of the country.

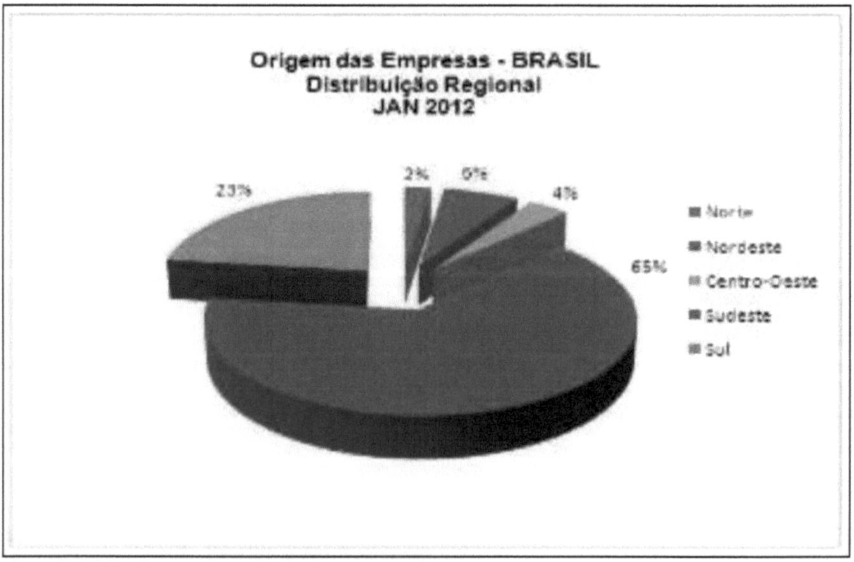

Figure 2. Origin of companies producing dangerous products
Source: http://ipr.dnit.gov.br/# (2018).

Continuing in the south/north direction, there are three points, one positive and two negative, that should be analysed. The positive point on the stretch is the electronic speed hump (figure 3) that controls the speed of vehicles travelling down the viaduct over Avenida Caxangá at 50 KM/h. This is a relevant point, as one of the biggest causes of traffic accidents is disrespect for speed limits. However, there is a CONTRAN resolution that recommends that electronic speed humps should not be signposted in front of the equipment, which can lead to sudden braking and even risky manoeuvres in front of such equipment, given the lack of respect by many vehicle drivers for the speed limits on the road.

Figure 3: Electronic speed bump in the south/north direction of BR 101

Source: Prepared by the authors (2018).

The second and third aspects observed are the poor maintenance of the lanes on the road both before and after the electronic speed bump, especially those in the centre and on the left where the flow of heavy vehicles is greatest, and the confluence on the right (Figure 5) located just after the speed bump, whose signage is precarious. The poor maintenance of the road can be seen in the unevenness (Figure 4), and even more so in the large number of potholes, especially in parts where the asphalt has been patched.

Figure 4: Gaps and potholes in the south/north direction
Figure 5: Confluence on the right after the hump.
Source: Prepared by the authors (2018).

As far as the north/south direction is concerned, there are far fewer products originating in the region to the north of the city of Recife, as can be seen from the survey carried out on the website of the Road Research Institute: only 5 types of product originate in the northern region.

The points observed in the north/south direction are worrying, given that the only positive point is the smaller number of products listed originating in the north of the city, but this should not be prioritised because the survey did not include the number of loads of these products transported, but rather the type, origin, destination and route followed.

The first negative point observed in the north/south direction is the Dois Irmãos viaduct, from where vehicles speed above the permitted speed, either because of the descent itself or because of the lack of signposting. It is also important to emphasise the lack of visibility of the confluence on the right just after descending from the viaduct, which is only warned by a poorly maintained sign covered by vegetation (Figure 6), and for those coming along the confluence there are no horizontal or vertical "give way" signs (Figure 7).

Figure 6. Confluence after the Dois Irmãos viaduct.
Figure 7. Lane confluent with the descent of the viaduct.
Source: Prepared by the authors (2018).

The second point to highlight is the heavy traffic on the shoulder by vehicles travelling to small businesses (Figure 8) just after descending the Dois Irmãos viaduct, as well as its poor state of repair. This situation conflicts with the Brazilian Traffic Code, which states that the shoulder is only used for emergency stops, putting traffic in the area at risk.

Figure 8. Disorganised occupation of plots bordering the carriageway
Source: Prepared by the authors (2018).

One of the adverse conditions that puts road users at risk is undoubtedly the poor maintenance of the carriageways. In the north/south direction, just after the Dois Irmãos viaduct, there are steep gradients (Figure 9) and a large number of potholes in this direction (Figure 10). These multi-factors, coupled with recklessness and the lack of signs indicating speed regulations, are the main causes of accidents.

Figura 9. Steep gradients in the road.

Figure 10. Potholes along the north/south stretch.

Source: Prepared by the authors (2018).

After these analyses, a comparison was made of the factors observed on site with item B of the section on Surveys of the Operational Conditions of the Road in the Manual for the Implementation of Emergency Action Plans to Respond to Accidents Involving the Road Transport of Dangerous Goods prepared by the DNIT, entitled "Field Inspection of the Road to Verify Operational Aspects". It contains sub-items with points that must be observed in order for traffic to fulfil minimum safety conditions:

- Condition of the road: the stretch showed some patches in the asphalt, unevenness and a large number of potholes, as seen in the analyses of the south/north and north/south directions, some of which were also caused by improperly made patches, which is a very high risk factor for accidents.

- Traffic profile: observations showed that traffic is heavy in the mornings and afternoons on weekdays, as well as on Saturday mornings. As far as the qualitative assessment is concerned, heavy goods traffic is heavy, compared to passenger vehicles, thus increasing the likelihood of accidents.

- Trespassing on the right of way: the right of way, according to the glossary, is the physical base on which a road is built, consisting of the carriageways, construction sites, road shoulders, signs and side safety lanes, up to the alignment of the fences separating the road from the marginal properties. In this respect, field observations found irregularities in the presence of street vendors (Figure 11) and a large flow of pedestrians and cyclists (Figure 12) from neighbouring communities.

Figure 11. Presence of street vendors on the roadside.
Figure 12. Flow of cyclists on the hard shoulder.
Source: Prepared by the authors (2018).

- Occurrence of stopped vehicles, abandoned vehicles and vehicles and/or loads outside the permitted standards: in this respect, the number of stopped vehicles is related to the drivers' search for products in the informal trade and in cases of emergency. The number of abandoned vehicles was non-existent. What was also observed were some vehicles travelling with their loads outside the established limits because they were exceeding the maximum width of the transport vehicle.
- Rubbish on the carriageway: on the carriageway itself, no rubbish was found that could affect traffic, but on the hard shoulder there was a large amount of rubbish of various kinds, such as plastic, wood, dead animals and so on. As a result, the possibility of this waste interfering with traffic is high, given that it is scattered on the carriageway without a suitable place to dispose of it.
- River crossings and unevenness in the carriageways, especially at the junctions with bridges over rivers with noble uses: this factor is the main point of the study, the junction of the BR 101 motorway with the Capibaribe River, in view of the risk of contamination of the river's waters. According

to the methodology applied by Viana (2009), the degree of risk at the points where the motorways cross the rivers is of high severity, while when the motorways cross the primary tributaries of these rivers, or even contributing rivers, this severity is considered medium, which demonstrates the environmental fragility of the stretch under study. As for the gaps in the seams of the bridge over the River Capibaribe, they are not as steep as those of the Dois Irmão viaduct, but they do require care, as the traffic is very intense.

- Segments with unused fenders; mainly bridges and river crossings: figure 13 shows the poor condition of the protection devices on the bridge, the railings are broken and there is no addition to the construction that could more efficiently protect both the river and the neighbouring population in the event of accidents.

Figure 13. Poor condition of the railings on the bridge over the Capibaribe.
Source: Prepared by the authors (2018).

- Loose animals on the right of way: no animals were seen on the carriageway on the days of the field visits, but there were loose animals on the land next to the motorway and also animals grazing on the roadbed that separates the motorway from the local carriageway (Figure 14), a factor that poses a high risk of accidents.

Figure 14. Presence of an animal in a side flowerbed.
Source: Prepared by the authors (2018).

-Fence fencing and state of repair of the side carriageway: there are small irregularities in the level of the carriageway, there is no drainage system and potholes, and the shrub cover on the neighbouring land at certain points is slightly impairing visibility due to the lack of proper maintenance.

Clandestine accesses and unsuitable junctions: in this respect, the stretch under study complies with safety standards, as there are no unsuitable accesses, junctions or U-turns. However, a worrying point is the recklessness of some motorbike drivers who use accesses intended for pedestrians as irregular U-turns (Figure 15).

Figure 15. Use of pedestrian access as a U-turn
Source: Prepared by the authors (2018).

- Drainage, run-off and areas of flooding in heavy rainfall: Figure 16 shows the precarious drainage system of the motorway. As well as containing a large amount of waste, there are no waste containment grids, nor even containment devices for chemical spills caused by road accidents involving the transport of dangerous products.

Figure 16. Poor drainage system.
Source: Prepared by the authors (2018).

- Existence of slopes at risk of slipping and other natural landforms: the risk of slipping can only be seen on the river bank near the head of the bridge due to the deforestation of riparian vegetation, as shown in figure 17. Irregular constructions are located on this stretch of bank.

Figure 17. Risk of slipping and/or erosion of the river bank with the carriageway.
Source: Prepared by the authors (2018).

- Guard rails: guard rails are physical elements with high resistance to vehicle impacts that are placed on the side of motorways to protect sensitive areas (DNIT, 2005). In this study, the sensitive area is characterised by the junction of BR 101 with the Capibaribe river, where they should be installed on both sides of the bridge with extensions that provide an effective level of safety.

The situation observed during the field visit was the precariousness of the infrastructure of these safety devices, as shown in figure 18: the side railings are built only on the slab of the bridge floor and are still in a poor state of repair. It is recommended that the length of the protection barriers should include the possible overhang, and extend for at least 30 metres before and after the existing side railings of the structures (DNIT, 2005).

Figura 18. Lack of protective barriers.
Source: Prepared by the authors (2018).

- Parking area: according to the ABNT - NBR - 14095 /1998 standard, car parks need to be away from urban areas, towns, schools, hospitals, water sources and environmental protection areas, and because of this it would not be possible to install these areas on the stretch under study in view of its proximity to the River Capibaribe.

- Specific signposting: according to the DNIT's General Signposting Plan, specific signposting for dangerous products should always be placed on the road in its critical stretches (DNIT, 2005), and nowhere in the area under study was any type of signposting for the guidance of dangerous cargo transporters observed.

As the Manual points out, places such as urban crossings, preservation areas and water sources, parking areas and places with restrictions on stopping, movement and speed, should be signposted to guide drivers of vehicles carrying dangerous loads. Figure 19 shows some examples of signs designed for this type of transport.

Sinalização de regulamentação | Área de proteção de manancial | Área de proteção da fauna e flora

Figura 19. Example of vertical signalling for the transport of dangerous goods.

Source: Prepared by the authors (2018).

CHAPTER 6

FINAL CONSIDERATIONS

The aim of this work was to analyse the structural conditions of BR 101 through an audit comparing the actual infrastructure situation with what is recommended in item B of the section on Surveys of the Operational Conditions of the Road, of the Manual for the Implementation of Emergency Action Plans for Responding to Accidents Involving the Road Transport of Dangerous Goods prepared by the DNIT, on the stretch between the neighbourhood of Dois Irmãos and the DETRAN community, in order to assess how susceptible this situation may be to accidents involving the transport of dangerous goods, so as to provide information for taking management measures, as well as drawing up an emergency response and contingency plan.

Identifying the dangerous products that are transported on the stretch under study showed a considerable diversity of products, with Class 8 products, corrosive substances, having the most records in the IPR system with a total of 14 different substances. In second place, those in Class 3, flammable liquids, had 12 different substances registered. It should always be noted that this research did not take into account the quantitative records of the products due to their unavailability.

With regard to the prognoses related to the impacts of registered products on the environment, more specifically the aquatic ecosystem, this research revealed that the vulnerability of the aquatic ecosystem is notorious when we look at the "ecological information" section of the MSDSs, because, of the 46 sheets surveyed, 34 emphasise the product's dangerousness in relation to the aquatic ecosystem, 10 advise avoiding the disposal of products in the aquatic environment and only 2 sheets, which refer to flammable gases, do not indicate any risks to the aquatic ecosystem.

In terms of analysing the environmental vulnerability of the Capibaribe River on the stretch under study, the infrastructure of the road does not offer safe conditions for those who use it as a route, especially when it comes to transporting dangerous products. The situation could be minimised by the installation of an electronic speed hump located before the junction of BR 101 with the watercourse in the south/north direction, given the reduced possibility of accidents due to the limited speed of vehicles travelling in this direction.

In view of this, the disagreement with the guidelines given by the DNIT itself and the norms established by other bodies can be seen in the vast majority of the items selected to ascertain the level of insecurity to which the Capibaribe River and the population bordering the BR 101 are subjected. This calls into question DNIT's ability to carry out its functions relating to the construction, maintenance, inspection and operation of the infrastructure of the segments of the Federal Road System, as an executing body within the Ministry of Transport.

Obviously, a more detailed quantitative analysis of environmental risks requires an assessment of the history of accidents involving the transport of dangerous goods in the region, as well as the development of modelling studies capable of simulating the dispersion of pollutants for each type of dangerous product, taking into account the peculiarities of the region.

Knowing the real scale of the problem requires detailed data, and this was not possible due to the fact that there was no response from the environmental organisations we contacted, such as the Brazilian Institute for the Environment and Renewable Natural Resources (IBAMA) and the State Environment Agency (CPRH), and from the Federal Highway Police.

The implementation of environmental emergency response plans is restricted to the National Plan for Prevention, Preparedness and Rapid Response to Environmental Emergencies with Dangerous Chemicals - P2R2, created by Federal Decree 5.098 of 2004, and there is no specific plan for the area under study, which would be ideal given its physical peculiarities.

Based on the questions raised for this work, it is clear that it will contribute to providing guidelines for environmental, safety, civil defence and civil society bodies in the implementation of a specific response plan for environmental accidents involving the transport of dangerous goods by road. It also contributes to the elaboration of measures aimed at developing a road transport system based on the environmental viability of the local road infrastructure due to the ecological, tourist and economic importance of the Capibaribe River for the city of Recife.

REFERENCES

ABIQUIM. Technical Department, Transport Committee. Manual for dealing with emergencies involving dangerous products. 4. ed. São Paulo: Atlas, 2011.

ANTT - National Land Transport Agency. Resolution no. 420 of 12 February 2004. Approves the Complementary Instructions to the Regulations for the Land Transport of Dangerous Goods.

ANTT - National Land Transport Agency. Provides general information about the organisation. Available at: <www.antt.gov.br>. Accessed on: 11 Feb. 2008.

ARAÚJO, Giovanni de Moraes de. Regulation of Land Transport of Dangerous Goods: Does Excessive Regulation Effectively Contribute to Reducing Accidents in the Transport of Dangerous Goods? Available at: <http://sobes.org.br /site /wp-content /uploads/2009/08/Prodperigosos.pdf> Accessed on: 27 Dec 2011.

ARAÚJO, Giovanni de Moraes de. Regulamentação do Transporte Terrestre de Produtos Perigosos Comentada. Rio de Janeiro: Gerenciamento Verde Editora, 2007. 964 p. ISBN 978-85-99337-13-2

BRAZILIAN ASSOCIATION OF TECHNICAL STANDARDS. NBR 14.001: Environmental Management System. Rio de Janeiro, 2004.

BASTOS, A. C. S.; ALMEIDA, J. R. Brazilian environmental licensing in the context of environmental impact assessment. In: Cunha; Guerra (org). Avaliação e perícia ambiental. 7. ed. Rio de Janeiro: Bertrand Brasil, 2006.

BRAZIL. Decree No. 96.044, of 18 May 1988. Approves the Regulations for the Road Transport of Dangerous Goods and makes other provisions. Official Gazette [of the Union], Brasília, 19 May 1988.

BRAZIL. Decree No. 2.657, of 3 July 1998, promulgating ILO Convention No. 170 concerning Safety in the Use of Chemicals at Work, signed in Geneva on 25 June 1990.

BRAZIL. National Department of Transport Infrastructure. Planning and Research Directorate. General Coordination of Studies and Research. Road Research Institute. Manual for implementing emergency action plans to deal with accidents involving the transport of dangerous goods by road. - Rio de Janeiro, 2005.

BRAZIL, Law 10.233, of 5 June 2001. Provides for the restructuring of waterway and land transport, creates the National Transport Policy Integration Council, the National Land Transport Agency, the National Waterway Transport Agency and the National Transport Infrastructure Department, and makes other provisions. 5 Jun 2001.

BRAZIL, Ministry of the Environment. Roadmap for drawing up an Emergency Action Plan-PAE . Available at: http://www.mma.gov.br/estruturas/sqa_p2r2_1/_arquivos/roteiro_pae.pdf. Accessed on: 27 January 2015.

CETESB. Chemical Products Manual. São Paulo, 2003.

CONAMA. CONAMA Resolutions, 1986 to 1991. Brasília: IBAMA, 1992.

DNIT. National Department of Transport Infrastructure. Provides general information about the institution. Available at: <www.dnit.gov.br>. Accessed on: 11 Feb. 2008.

DNIT. Road Research Institute. Manual for the Implementation of Emergency Action Plans for Responding to Accidents Involving the Road Transport of Dangerous Goods. Publication 716. Rio de Janeiro 2005.

EMBRAPA, 2010. Diagnosis and alternatives for the environmental recovery of the Guandu River Basin (BHRG) - RJ, 2010.

FERREIRA, Marco Antônio. Types of Modals. Available at:<http://www.ietec. com.br /site/techoje/categoria/detalhe_artigo/670>. Accessed on: 20 December 2011.

FRICKE, G. T. Vulnerability of the campinas water supply system to accidents involving the transport of dangerous products. Master's dissertation. Faculty of Civil Engineering / UNICAMP, 1992.

IPR. Available at: <http://ipr.dnit.gov.br/>. Accessed on 19 Dec 2011.

MELO, J.G. da SILVA; OLIVEIRA, T.H.;DA SILVA, C.A.; TORRES, M.F.A.; GALVÍNCIO, J.D.; DA SILVA, H.A. Space-time analysis of the mangrove swamp of the lower course of the Capibaribe River, Recife - PE: a contribution to environmental management. Anais XV Simpósio Brasileiro de Sensoriamento Remoto - SBSR, Curitiba, PR, Brazil, 30 April to 05 May 2011, INPE p.6586.

MINISTRY OF JUSTICE. Hazardous Products Emergency Intervention Course: SENASP/EAD. Brasília. 2008.

available at :
<http://portal.mj.gov.br/data/Pages/MJ27337B92ITEMID3CF184807C2B4431A47ACDF5C 6426B93PTBRNN.htm>. Accessed on 21/12/2011

MORGADO, C. R. V and HADDAD, A. N. Elementos de Segurança Ambiental. 1. ed. Rio de Janeiro: Bio-Rio Foundation, 2002.

INTERNATIONAL LABOUR ORGANISATION. ILO 174: Prevention of Major Industrial Accidents Convention. Geneva, 1993.

OCCUPATION HEALTH AND SAFETY MANAGEMENT SYSTEMS. OHSAS 18.001: Occupational Health and Safety Management System. United Kingdom, 2007.

SÁNCHEZ, Luis Henrique. Environmental Impact Assessment: concepts and methods. São Paulo: Oficina Textos, 2006. 495 p.ISBN 85-86238-59-7

SCHAEFFER-NOVELLI, Y. Group of Ecosystems: Mangrove, Marisma and Apicum. São Paulo. Brazil, 2001

SILVA,E. R. A. C.Potential for hydro-agricultural development in the Pontal Creek - PE: subsidies for the intelligent use of water. Dissertation (Master's in Geography) - UFPE, Federal University of Pernambuco. Recife, 2014.

VIANA, Viviane Jupiassú: Environmental Risks Associated with the Transport of Dangerous Products in the Area of Influence of the Guandu WTP/RJ. Dissertation (Master's in Environmental Engineering) - UERJ, University of Rio de Janeiro, Rio de Janeiro, 2009.

More Books!

yes

I want morebooks!

Buy your books fast and straightforward online - at one of world's fastest growing online book stores! Environmentally sound due to Print-on-Demand technologies.

Buy your books online at

www.morebooks.shop

Kaufen Sie Ihre Bücher schnell und unkompliziert online – auf einer der am schnellsten wachsenden Buchhandelsplattformen weltweit! Dank Print-On-Demand umwelt- und ressourcenschonend produzi ert.

Bücher schneller online kaufen

www.morebooks.shop

info@omniscriptum.com
www.omniscriptum.com

OMNIScriptum

MIX
Papier aus verantwortungsvollen Quellen
Paper from responsible sources
FSC® C105338

Printed by Books on Demand GmbH, Norderstedt / Germany